YOUR KNOWLEDGE HAS VALUE

AF144203

Jonas Sauer

Aus der Reihe: e-fellows.net stipendiaten-wissen

e-fellows.net (Hrsg.)

Band 245

Analysis of the Landau Solution

GRIN Verlag

Bibliografische Information der Deutschen Nationalbibliothek:

Die Deutsche Bibliothek verzeichnet diese Publikation in der Deutschen National-
bibliografie; detaillierte bibliografische Daten sind im Internet über http://dnb.d-
nb.de/ abrufbar.

Imprint:

Copyright © 2010 GRIN Verlag GmbH
Druck und Bindung: Books on Demand GmbH, Norderstedt Germany
ISBN: 978-3-656-01637-3

GRIN - Your knowledge has value

Der GRIN Verlag publiziert seit 1998 wissenschaftliche Arbeiten von Studenten, Hochschullehrern und anderen Akademikern als eBook und gedrucktes Buch. Die Verlagswebsite www.grin.com ist die ideale Plattform zur Veröffentlichung von Hausarbeiten, Abschlussarbeiten, wissenschaftlichen Aufsätzen, Dissertationen und Fachbüchern.

Visit us on the internet:

http://www.grin.com/

http://www.facebook.com/grincom

http://www.twitter.com/grin_com

Analysis of the Landau Solution

Bachelor-Thesis von Jonas Sauer

TECHNISCHE
UNIVERSITÄT
DARMSTADT

Fachbereich Mathematik
Forschungsbereich Analysis

Analysis of the Landau Solution

Vorgelegte Bachelor-Thesis von Jonas Sauer

1. Gutachten: Prof. Dr. Reinhard Farwig
2. Gutachten: Prof. Dr. Dr. h.c. Hans-Dieter Alber

Tag der Einreichung:

Very special thanks to Prof. Dr. Farwig from the Technische Universität Darmstadt, who sparked my interest in the topic of this present paper and who provided a lot of helpful corrections and hints.
I also want to thank Prof. RNDr. Josef Málek, CSc., DSc. from Univerzita Karlova v Praze for patronising me during my stay in Prague.

Prague and Darmstadt, August 2010

Contents

1 Introduction

In 1944 Lev D. Landau [15] found a non-trivial solution to a stationary Navier-Stokes flow on $\mathbb{R}^3\backslash\{0\}$, which was symmetric around some axis and fulfilled the condition, that the velocity decayed linearly and the pressure quadratically in $|x|$. If the axis of symmetry is the x_3-axis, the solution is of the form

$$
\begin{aligned}
u_1(x) &= 2d\frac{x_1\left(x_3 - d\,|x|\right)}{|x|\left(|x| - dx_3\right)^2}, \\
u_2(x) &= 2d\frac{x_2\left(x_3 - d\,|x|\right)}{|x|\left(|x| - dx_3\right)^2}, \\
u_3(x) &= 2d\frac{|x|^2 - 2dx_3\,|x| + x_3^2}{|x|\left(|x| - dx_3\right)^2}, \\
p(x) &= 4d\frac{x_3 - d\,|x|}{|x|\left(|x| - dx_3\right)^2},
\end{aligned}
\tag{1.1}
$$

where $d \in (-1, 1)$. We note here, that Squire [19] found this solution independently in 1951. Landau was examining this flow model when considering the following physical experiment. Imagine a thin pipe in a fluid without any boundary conditions. The fluid is driven by a small jet outflowing the pipe. If one looks at the limit case, where the diameter of the pipe shrinks to zero, there is no force acting on the fluid on $\mathbb{R}^3\backslash\{0\}$. Nevertheless, if one wants to describe the physical behaviour of the fluid completely, one has to take into account the force acting at the origin, i.e. at the orifice of the pipe. Therefore, one has to extend the solution in a suitable manner to the whole space $\mathbb{R}^3$. This can be done if one admits also weak solutions in the space of distributions which satisfy

$$
\begin{aligned}
\operatorname{div} u &= 0, \\
-\Delta u + \nabla p + u \cdot \nabla u &= b\delta,
\end{aligned}
\tag{1.2}
$$

where b is a vector and δ is the Dirac distribution.

When looking at the solutions found by Landau, it seems quite natural to ask, if there are also reasonable physical interpretations for the case $|d| \geq 1$. Therefore we first derive Landau's solution with the assumptions posed in the beginning, following an ansatz by Batchelor [3]. From there it will be clear, that the velocity is unbounded on a cone, if $|d| > 1$. As it turns out, such a modified Landau solution is no longer a solution to a Navier-Stokes system - clearly not in the classical, but as well neither in the weak nor very weak sense. If $d = \pm 1$, the velocity will be unbounded on the half line $\{x \in \mathbb{R}^3 : x_1 = x_2 = 0, \pm x_3 \geq 0\}$. In this case, we have a bit more insight in the behaviour of the modified Landau solution, yet still no physically reasonable interpretation can be given in the whole space $\mathbb{R}^3$. In any case, one can consider modified Landau solutions to be solutions in certain subsets of $\mathbb{R}^3$, where they yield some interesting streamline plots, which are also provided in this thesis. Furthermore, we give an overview over known results concerning the Landau solution.

In this paper we will use some common notation. Ω usually denotes an open subset of $\mathbb{R}^3$. For the space of k-times continuously differentiable functions from Ω to $\mathbb{R}^n$ we write $C^k(\Omega, \mathbb{R}^n)$. Furthermore we set $C^k(\Omega, \mathbb{R}) = C^k(\Omega)$ and analogously for the other spaces introduced here. We will also not distinguish between $\left(C^k(\Omega)\right)^3$ and $C^k(\Omega)$. The space of infinitely often differentiable functions with compact support, i.e. the test functions, will be denoted by $C_0^\infty(\Omega)$ or sometimes by $\mathscr{D}(\Omega)$. Note, that its dual space

$\mathscr{D}'(\Omega)$ is the space of distributions. $L^p(\Omega)$ is the Lebesgue space of p-integrable functions and $W^{k,p}(\Omega)$ is the Sobolev space of k-times weakly differentiable p-integrable functions, whose derivatives are also p-integrable. The subscript loc indicates, that the functions need to be p-integrable only on every compact subset of Ω. A ball around the origin with radius R will be denoted by B_R.

Due to the concepts dealt with in this paper, we will also use some notation from fluid dynamics. The tensor product of two vectors $\boldsymbol{a}$ and $\boldsymbol{b}$ is the 3×3-matrix $\boldsymbol{a} \otimes \boldsymbol{b} := (a_i b_j)_{ij}$. The Cauchy stress tensor will be denoted by $T = -p\mathbb{I} + \nabla u + (\nabla u)^T$. Note that the momentum flux density tensor is given by $u \otimes u - T$. The reader should note, that throughout this paper we will work without explicitly taking the dimension of physical quantities into account. Furthermore we set for the density of the fluid $\rho = 1$. Except for the solution u and the position vector x, vectors in $\mathbb{R}^3$ and vector-valued functions will be denoted by a bold face letter.

When speaking about distributions, the angled brackets denote the action of a distribution on a test function, i.e. $\langle f, \varphi \rangle$ means, that the distribution f is acting on the test function φ. Unless explicitly said otherwise, the symbol δ will always represent the Dirac distribution.

In the appendix the reader may find some basic theorems used in this paper. We usually refrain from referring explicitly to the theorems provided in the appendix, as they are mostly well-known and only meant for the convenience of the reader.

2 Preliminaries

As we will be concerned with different kinds of solutions to a Navier-Stokes system, we first have to introduce the setup we are working with.

Following the usual procedures in fluid dynamics, we will work with quantities such as the velocity u or the pressure p in order to describe certain flows. These quantities are functions of space and time, that is their domain is

$$\mathscr{M} := \{(x,t) \in \Omega_t \times (0,T)\} \subset \mathbb{R}^4, \quad 0 < T < \infty \tag{2.1}$$

where Ω_t is the (open) domain occupied by the fluid at time t. We will assume, that $\mathscr{M}$ is open. A sufficient condition for this is, that Ω_t changes continuously in time. In fact, we will be only considering the situation, where $\Omega_t = \Omega$ will be constant. Nevertheless, in this section we will state the main results with explicitly taking the change in time into account.

2.1 The Transport Theorem

Our mathematical model will treat fluids as a continuum, and our fundamental hypothesis is, that exactly one fluid particle passes through any point $x \in \Omega$ at any time $t \in (0,T)$.

That gives rise to some results laid out in this section. The first is the Transport Theorem, which is of utmost importance. Let us consider a system of fluid particles in a bounded domain $\mathscr{V}(t) \subset \Omega_t$ at a time t. It is well known, that for given velocity $u \in C^1(\mathscr{M})$, the initial value problem

$$\frac{dx}{dt} = u,$$
$$x(t_0) = X \tag{2.2}$$

has a unique maximal solution $x(t) = \varphi(X, t_0; t)$, where φ has continuous first derivatives and both $\frac{\partial \varphi}{\partial t_0}$ and $\frac{\partial \varphi}{\partial t}$ have continuous first spatial derivatives ([14], theorem 10.1.1, 11.1.5, 13.1.1). If we fix t_0, φ defines the change of the domain $\mathscr{V}(t)$ with time

$$\mathscr{V}(t) = \{\varphi(X, t_0; t) : X \in \mathscr{V}(t_0)\}. \tag{2.3}$$

We are now ready to state the assertion of the Transport Theorem.

Transport Theorem. *Let $t_0 \in (0,T)$ and $\mathscr{V}(t_0)$ be a bounded domain with $\overline{\mathscr{V}(t_0)} \subset \Omega_{t_0}$. Let $\varphi : \mathscr{V}(t_0) \rightarrow \mathscr{V}(t)$ be a continuously differentiable and bijective map (defining the change of $\mathscr{V}(t_0)$ with time) with continuous and bounded Jacobian determinant J, which satisfies $J(X,t) > 0$ for all $X \in \mathscr{V}(t_0)$ and for all t in an interval $(t_1, t_2) \in (0,T)$ containing t_0. Let $F : \mathscr{M} \rightarrow \mathbb{R}$ have continuous and bounded first derivatives on the set $\{(x,t) : t \in (t_1, t_2), x \in \mathscr{V}(t)\}$.*
Then we have for all $t \in (t_1, t_2)$ that there exists the finite derivative

$$\frac{d}{dt} \int_{\mathscr{V}(t)} F(x,t) \, dV = \int_{\mathscr{V}(t)} \left(\frac{\partial F}{\partial t}(x,t) + \text{div}\,(Fu)(x,t) \right) dV. \tag{2.4}$$

Proof. Observe, that there exists an interval (t_1, t_2) containing t_0, such that φ and J satisfy the conditions of the theorem, see for example [10]. Then, by the Change of Variable Formula, we get

$$\frac{d}{dt} \int_{\mathcal{V}(t)} F(x,t)\, dV(x) = \frac{d}{dt} \int_{\mathcal{V}(t_0)} F(\varphi(X,t_0;t)) J(X,t)\, dV(X). \tag{2.5}$$

By a straight forward calculation which can be found in [9], we see that $\frac{\partial J}{\partial t}(X,t) = J(X,t)\operatorname{div} u(X,t)$. If one calls the integrand appearing in the right integral $\overline{F}(X,t)$, we see that for all t it holds that $\overline{F}$ is measurable as a function of X and that it has finite derivatives for almost every $X \in \mathcal{V}(t_0)$. Furthermore, the time derivatives of $\overline{F}$ are bounded by an integrable majorizing function g and there exists $t^* \in (t_1, t_2)$ such that $\overline{F}(X, t^*)$ is integrable. We therefore get by the theorem on differentiation of an integral with respect to one parameter

$$\frac{d}{dt} \int_{\mathcal{V}(t_0)} \overline{F}\, dV(X) = \int_{\mathcal{V}(t_0)} \left(\left(\frac{\partial F}{\partial t} + \sum_{j=1}^{3} \frac{\partial F}{\partial x_j} \frac{\partial \varphi_j}{\partial t} \right) J + F J \operatorname{div} u \right) dV(X) \tag{2.6}$$

$$= \int_{\mathcal{V}(t_0)} \left(\frac{\partial F}{\partial t} + \sum_{j=1}^{3} \frac{\partial F}{\partial x_j} u_j + F \operatorname{div} u \right) J\, dV(X) \tag{2.7}$$

$$= \int_{\mathcal{V}(t)} \left(\frac{\partial F}{\partial t} + \nabla F \cdot u + F \operatorname{div} u \right) dV(x) \tag{2.8}$$

$$= \int_{\mathcal{V}(t)} \left(\frac{\partial F}{\partial t} + \operatorname{div}(Fu) \right) dV(x). \tag{2.9}$$

$$\tag{2.10}$$

Please observe, that we omitted the arguments here and that we used the well-known relation $\nabla F \cdot u + F \operatorname{div} u = \operatorname{div}(Fu)$. □

Using this theorem, we can derive the mathematical formulation of the fundamental physical laws of conservation, such as the conservation of mass or the conservation of momentum, in differential form, known as the governing equations of fluid dynamics.

2.2 Governing equations of fluid dynamics

Let $\rho : \mathcal{M} \to (0, \infty)$ denote the density of the fluid. Then by the conservation law of mass and by the Transport Theorem, we get

$$0 = \frac{d}{dt} \int_{\mathcal{V}} \rho(x,t)\, dV = \int_{\mathcal{V}} \left(\frac{\partial \rho}{\partial t}(x,t) + \operatorname{div}(\rho u)(x,t) \right) dV, \tag{2.11}$$

for all control volumes $\mathcal{V}(t) \subset \overline{\mathcal{V}(t)} \subset \Omega_t$. By the continuity of the integrand we thus conclude

$$\frac{\partial \rho}{\partial t} + \operatorname{div}(\rho u) = 0 \quad \text{in } \mathcal{M}. \tag{2.12}$$

This equation is known as the **continuity equation**. In the steady-state situation, that is, if all time derivatives vanish, this simplifies to $\operatorname{div}(\rho u) = 0$. Throughout this paper we will only be concerned with the case, where the density is constant also with respect to the spatial components, i.e. with incompressible flows. Obviously, in that case the conservation law of mass is turned into the statement

$$\operatorname{div} u = 0. \tag{2.13}$$

So we see, that there is no flux of mass across any closed surface for incompressible fluids, or in other words: At any time and for any control volume, the inward flow equals the outward flow.

In the same manner one can examine the conservation law of momentum. Recall that the force is defined as the time derivative of the momentum. Thus, for the total force $\mathbf{F}$ acting on a domain $\mathcal{V}(t) \in \Omega_t$, we have the relation

$$\mathbf{F}(\mathcal{V}(t), t) = \frac{\partial \mathbf{M}}{\partial t}(\mathcal{V}(t), t) = \frac{d}{dt} \int_{\mathcal{V}(t)} \rho u \, dV \qquad (2.14)$$

$$= \int_{\mathcal{V}(t)} \left(\frac{\partial (\rho u)}{\partial t} + \mathrm{div}\,(\rho u \otimes u) \right) dV. \qquad (2.15)$$

But on the other hand, for Newtonian fluids we can decompose the total force $\mathbf{F}$ into the volume force and the surface force by virtue of its density f and the Cauchy stress tensor T

$$\mathbf{F}(\mathcal{V}(t), t) = \int_{\mathcal{V}(t)} \rho f \, dV + \int_{\partial \mathcal{V}(t)} T n \, dS. \qquad (2.16)$$

Combining these two expressions for the total force and using Green's theorem will yield the equations of motion

$$\frac{\partial \rho u}{\partial t} + \mathrm{div}\,(\rho u \otimes u) = \rho f + \mathrm{div}\,T = \rho f - \nabla p + \nabla(\mathrm{div}\,u) + \mathrm{div}\left(\nabla u + (\nabla u)^T\right). \qquad (2.17)$$

Observe, that we have the two relations

$$\mathrm{div}\left(\nabla u + (\nabla u)^T\right) = \Delta u + \nabla \mathrm{div}\,u, \quad \frac{\partial \rho u}{\partial t} + \mathrm{div}\,(\rho u \otimes u) = \rho \left(\frac{\partial u}{\partial t} + (u \cdot \nabla)u \right). \qquad (2.18)$$

We use here the convention

$$u \cdot \nabla u = (u \cdot \nabla)u = \sum_{j=1}^{3} u_j \frac{\partial u}{\partial x_j}. \qquad (2.19)$$

This leads in the stationary case, i.e. $\mathrm{div}\,u = 0$ to the Navier-Stokes equation

$$-\Delta u + \nabla p + u \cdot \nabla u = f. \qquad (2.20)$$

Usually we refer to the system of equations

$$\begin{aligned} \mathrm{div}\,u &= 0, \\ -\Delta u + \nabla p + u \cdot \nabla u &= f, \end{aligned} \qquad (2.21)$$

as the **Navier-Stokes system with force** f.

2.3 Concepts of a solution

In the classical work frame, we are usually interested in solutions to partial differential equations, which are at least as smooth as the degree of the equation. This concept of a solution turns out to be not suitable for certain applications. There we have to use a more general definition of a solution to a partial differential equation, which will allow us to give a physical interpretation to certain equations. First let us state the classical definition.

Definition 2.3.1. Let $n, m \in \mathbb{N}$ and let $\Omega \subset \mathbb{R}^n$. Then we call $u : \Omega \to \mathbb{R}^m$ a **classical solution** on Ω to the partial differential equation

$$F\left(\xi, \{D^\alpha \psi(\xi) : |\alpha| \leq k\}\right) = 0, \tag{2.22}$$

where ξ is an independent variable, if $u \in C^k(\Omega)$ and for all $x \in \Omega$ we have $F(x, \{D^\alpha u(x) : |\alpha| \leq k\}) = 0$.

As already mentioned, this concept of a solution to partial differential equations is not always an appropriate approach. There are various concepts of solutions, but we will state here two definitions, which we will be needing when analysing the behaviour of the Landau solution.

Definition 2.3.2. Let $n \in \mathbb{N}$ and let $\Omega \subset \mathbb{R}^n$ and let L be a linear differential operator. Then we call $u \in \mathscr{D}'(\Omega)$ a **weak solution in the sense of distributions** or short **weak solution** on Ω to the partial differential equation

$$L\psi = 0, \tag{2.23}$$

if for all test functions $\varphi \in C_0^\infty(\Omega)$ it holds true that

$$\langle Lu, \varphi \rangle = 0. \tag{2.24}$$

This definition extends in an obvious way to solutions to partial differential equations in multiple functions.

Observe, that for distributions we have the definition $\langle D^\alpha f, \varphi \rangle = (-1)^{|\alpha|} \langle f, D^\alpha \varphi \rangle$. Furthermore, if a function f is in $L^1_{loc}(\Omega)$, we may regard it as a distribution by setting

$$\langle f, \varphi \rangle = \int_\Omega f\varphi \, dV. \tag{2.25}$$

Especially for the case of the Navier-Stokes system, we may consider the concept of a very weak solution in the following sense. For this definition we have to recall, that if $\Omega \subset \mathbb{R}^3$ is bounded, then $L^1_{loc}(\Omega) \supset L^2_{loc}(\Omega)$.

Definition 2.3.3. Let $f \in \mathscr{D}'(\Omega)$, where Ω is a finite region in $\mathbb{R}^3$. Then $u : \Omega \to \mathbb{R}^3$ is a **very weak solution** on Ω to the Navier-Stokes system (2.21) with force f, if $u \in L^2_{loc}(\Omega)$ and it holds true that u is weakly divergence-free in the sense of distributions as defined above and for all divergence-free test functions $\varphi \in C_0^\infty(\Omega, \mathbb{R}^3)$ we have

$$\int_\Omega \left(-u \cdot \Delta\varphi - u \cdot (\nabla\varphi)\, u\right) dV = \langle f, \varphi \rangle. \tag{2.26}$$

Note that due the special properties of φ, namely $\operatorname{div} \varphi = 0$, we do not have to take the pressure p of the original Navier-Stokes equation into account.

3 Analysis of the classical case

3.1 The classical Landau solution

Theorem 3.1.1. *Let $d \in (-1,1)$. Then on $\mathbb{R}^3 \setminus \{0\}$ a smooth solution (u,p) to the Navier-Stokes system (2.21) which is symmetric, but does not rotate about the x_3-axis and fulfils the homogeneity conditions $u(x) = \frac{1}{|x|}u(\frac{x}{|x|})$ and $p(x) = \frac{1}{|x|^2}p(\frac{x}{|x|})$ is given by (1.1). Such a solution is called a* **Landau solution**. *Furthermore, every smooth solution satisfying these conditions is a Landau solution.*

Rather than checking all the conditions, we will derive the solution from the scratch to have a better insight in the physical meanings of the respective terms, which will help us understanding the more general situation later.

The main idea of our attempt to find a solution to the Navier-Stokes system is, that we reduce the problem to a solvable ordinary differential equation in only one variable. Even though the axisymmetry of the system might suggest cylindrical coordinates, it turns out that using spherical coordinates yields the fastest approach towards this, for then we can make use of our knowledge about the radial dependence of our solution. The proof provided here is due to [3] and [19].

Proof. Since the Navier-Stokes flow is assumed to be stationary and has, if one assumes in advance no rotation of the fluid around its axis of symmetry, only two degrees of freedom, the continuity equation is identically fulfilled by finding a stream function ψ as introduced in [3], such that the velocity components in spherical coordinates are described by

$$u^{(r)} = \frac{1}{r^2 \sin\theta} \frac{\partial \psi}{\partial \theta}, \quad u^{(\theta)} = -\frac{1}{r \sin\theta} \frac{\partial \psi}{\partial r}, \quad u^{(\phi)} = 0. \tag{3.1}$$

In view of the posed problem we assume, that the velocity is (-1)-homogeneous. That means by (3.1), that there exists a function f, such that

$$\psi(r,\theta) = r f(\theta). \tag{3.2}$$

Similarly, as the pressure p is supposed to be (-2)-homogeneous, we can introduce a function g such that

$$p = \frac{1}{r^2} g(\theta). \tag{3.3}$$

If one takes the Navier-Stokes equation in the form for spherical coordinates and plugs in (3.1), (3.2) and (3.3), one gets

$$\frac{1}{r^3}\left(\left(\frac{f_\theta}{\eta}\right)^2 + \frac{f}{\eta}\left(\frac{f_\theta}{\eta}\right)_\theta + \frac{f^2}{\eta^2}\right) = -\frac{2}{r^3}g - \frac{1}{r^3}\left(\frac{1}{\eta}\left(\eta\left(\frac{f_\theta}{\eta}\right)\right)_\theta\right)_\theta \underbrace{-\frac{2f_\theta}{\eta} + 2\left(\frac{f}{\eta}\right)_\theta + \frac{2f\xi}{\eta^2}}_{=0}, \tag{3.4}$$

$$\frac{1}{r^3}\frac{f}{\eta}\left(\frac{f}{\eta}\right)_\theta = -\frac{1}{r^3}g_\theta + \frac{1}{r^3}\left(-\frac{1}{\eta}\left(\eta\left(\frac{f}{\eta}\right)\right)_\theta\right)_\theta + 2\left(\frac{f_\theta}{\eta}\right)_\theta + \frac{f}{\eta^3}\right), \tag{3.5}$$

---

where we have set $\eta := \sin\theta$ and $\xi := \cos\theta$.

As one can see, this eliminates the variable r from the equations. Noting, that

$$\frac{f_\theta}{\eta} = -f_\xi$$

one thereby receives the equations

$$
\begin{aligned}
-2g &= \left(\frac{f_\theta}{\eta}\right)^2 + \frac{f}{\eta}\left(\frac{f_\theta}{\eta}\right)_\theta + \frac{f^2}{\eta^2} + \frac{1}{\eta}\left(\eta\left(\frac{f_\theta}{\eta}\right)\right)_\theta, \\
&= (f_\xi)^2 - \frac{f}{\eta}(f_\xi)_\theta + \frac{f^2}{1-\xi^2} + \frac{1}{\eta}\left(\eta^2\frac{(-f_\xi)_\theta}{\eta}\right)_\theta, \\
&= (f_\xi)^2 + f f_{\xi\xi} + \frac{f^2}{1-\xi^2} - \left(\left(1-\xi^2\right)f_{\xi\xi}\right)_\xi \\
&= \left(f f_\xi - \left(1-\xi^2\right)f_{\xi\xi}\right)_\xi + \frac{f^2}{1-\xi^2}.
\end{aligned}
\tag{3.6}
$$

$$
\begin{aligned}
-g_\xi &= \frac{1}{\eta}\left[-\frac{f}{\eta}\left(\frac{f}{\eta}\right)_\theta - \frac{1}{\eta}\left(\eta\left(\frac{f}{\eta}\right)\right)_{\theta\theta} + 2\left(\frac{f_\theta}{\eta}\right)_\theta + \frac{f}{\eta^3}\right] \\
&= \frac{1}{\eta}\left[-\frac{f}{\eta}\left(\frac{f_\theta\eta - f\xi}{\eta^2}\right) - \frac{1}{\eta}\left(\frac{f_\theta\eta - f\xi}{\eta}\right)_\theta - 2(f_\xi)_\theta + \frac{f}{\eta^3}\right] \\
&= \frac{f}{\eta^2}\left(f_\xi + \frac{f\xi}{\eta^2}\right) - \frac{1}{\eta^2}\left(f_{\theta\theta} - \frac{f_\theta\xi}{\eta}\right) + 2f_{\xi\xi} \\
&= \frac{1}{2}\left(\frac{f^2}{1-\xi^2}\right)_\xi + f_{\xi\xi}.
\end{aligned}
\tag{3.7}
$$

By integrating the last equation we therefore arrive at the identity

$$2f_\xi + c_1 = \left(f f_\xi - \left(1-\xi^2\right)f_{\xi\xi}\right)_\xi,\tag{3.8}$$

where c_1 is an integration constant. Integrating two more times, one gets respectively (where c_2 and c_3 are again constants of integration)

$$f f_\xi - \left(1-\xi^2\right)f_{\xi\xi} - 2f = c_1\xi + c_2,\tag{3.9}$$

$$\frac{1}{2}f^2 - \left(1-\xi^2\right)f_\xi - 2\xi f = c_1\xi^2 + c_2\xi + c_3.\tag{3.10}$$

So far, we have achieved our goal of transforming our problem to an ordinary differential equation, yet we still have to choose the constants in a way such that the requirements of the system (2.21) are matched. As u is required to be a smooth solution and due to the axisymmetry, $u^{(\theta)}$ must vanish for $\theta = 0$ and $\theta = \pi$, i.e. for $\xi = 1$ and $\xi = -1$. From (3.1) we get, that there is also f has to vanish. Plugging those two values in (3.10) and substracting the resulting equations from each other, we get $c_2 = 0$. By adding them, we get $c_1 + c_3 = 0$. But analogously, we get from (3.9), that $c_1 = 0$. This leaves us with the only possible choice $c_1 = c_2 = c_3 = 0$ for the constants.

Setting $f(\xi) = (1 - \xi^2)h(\xi)$ leads to an ordinary differential equation $h^2 - 2h_\xi = 0$. Its solution on $\xi \in [-1, 1]$ ($\theta \in [0, \pi]$) is

$$h(\xi) = \frac{2d}{1-d\xi},$$

where $|d| < 1$, as can be derived easily by seperation of variables. Therefore, we get the solution for f

$$f(\theta) = \frac{2d(\sin^2 \theta)}{1 - d \cos \theta}. \tag{3.11}$$

Plugging in this result in the expressions of (3.1) and (3.6), we are able to compute the solution in spherical coordinates

$$u^{(r)} = \frac{1}{r} \left(\frac{4d \cos \theta}{1 - d \cos \theta} - \frac{2d^2 \sin^2 \theta}{(1 - d \cos \theta)^2} \right), \quad u^{(\theta)} = -\frac{2d}{r} \left(\frac{\sin \theta}{1 - d \cos \theta} \right), \quad p = \frac{4d}{r^2} \left(\frac{\cos \theta - d}{(1 - d \cos \theta)^2} \right). \tag{3.12}$$

As the axis of symmetry is the x_3-axis, we have that

$$\begin{aligned}
x_1 &= r \sin \theta \cos \phi, \\
x_2 &= r \sin \theta \sin \phi, \\
x_3 &= r \cos \theta.
\end{aligned} \tag{3.13}$$

Therefore the solutions transform into cartesian coordinates as follows

$$\begin{aligned}
u_1 &= u^{(r)} \sin \theta \cos \phi + u^{(\theta)} \cos \theta \cos \phi \\
&= \frac{1}{|x|} \left(\frac{4d x_3 x_1}{|x|(|x| - dx_3)} - \frac{2d^2(|x|^2 - x_3^2)x_2}{|x|(|x| - dx_3)^2} - \frac{2d x_3 x_1}{|x|(|x| - dx_3)} \right) \\
&= 2d \frac{x_1}{|x|} \left(\frac{x_3(|x| - dx_3) - d(|x|^2 - x_3^2)}{|x|(|x| - dx_3)^2} \right) \\
&= 2d \frac{x_1(x_3 - d|x|)}{|x|(|x| - dx_3)^2}
\end{aligned} \tag{3.14}$$

$$\begin{aligned}
u_2 &= u^{(r)} \sin \theta \sin \phi + u^{(\theta)} \cos \theta \sin \phi \\
&= \frac{1}{|x|} \left(\frac{4d x_3 x_2}{|x|(|x| - dx_3)} - \frac{2d^2(|x|^2 - x_3^2)x_2}{|x|(|x| - dx_3)^2} - \frac{2d x_3 x_2}{|x|(|x| - dx_3)} \right) \\
&= 2d \frac{x_2}{|x|} \left(\frac{x_3(|x| - dx_3) - d(|x|^2 - x_3^2)}{|x|(|x| - dx_3)^2} \right) \\
&= 2d \frac{x_2(x_3 - d|x|)}{|x|(|x| - dx_3)^2}
\end{aligned} \tag{3.15}$$

$$\begin{aligned}
u_3 &= u^{(r)} \cos \theta - u^{(\theta)} \sin \theta \\
&= \frac{1}{|x|} \left(\frac{4d x_3^2}{|x|(|x| - dx_3)} - \frac{2d^2(|x|^2 - x_3^2)x_3}{|x|(|x| - dx_3)^2} + \frac{2d(|x|^2 - x_3^2)}{|x|(|x| - dx_3)} \right) \\
&= 2d \frac{1}{|x|} \left(\frac{x_3^2(|x| - dx_3) - d(|x|^2 - x_3^2)x_3 + |x|^2(|x| - dx_3)}{|x|(|x| - dx_3)^2} \right) \\
&= 2d \frac{|x|^2 - 2d x_3 |x| + x_3^2}{|x|(|x| - dx_3)^2}
\end{aligned} \tag{3.16}$$

$$p = 4d \frac{x_3 - d|x|}{|x|(|x| - dx_3)^2}. \tag{3.17}$$

$\square$

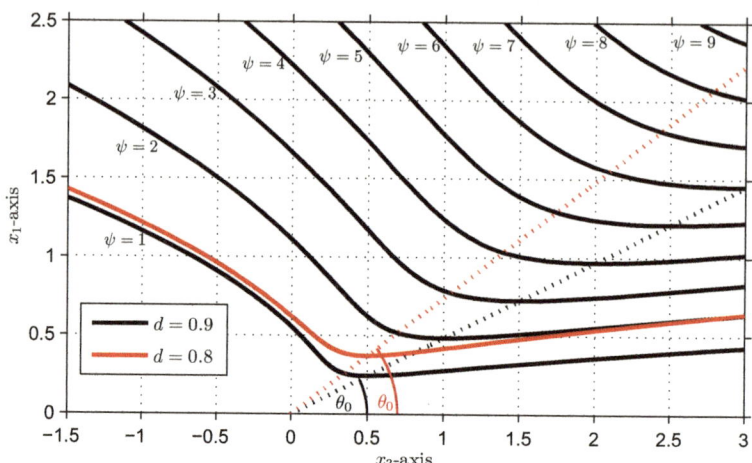

Figure 3.1.: Streamlines of the flow of a Landau solution for $d = 0.8$ with $\psi = 1$ and for $d = 0.9$ with different values for ψ.

Remark. In figure 3.1 the flow of this solution is plotted for $d = 0.8$ and $d = .9$. Since the flow is axisymmetric, we can understand the behaviour of the whole flow by just looking at the case $\phi = 0$. As one can see, each streamline (or more precisely streamtube) has a unique area, called the **throat of the streamtube**, where it is nearest to the axis of symmetry. Since ψ is constant on each streamline, r is proportional to $\frac{1}{f}$ and we thus have for the distance between streamtube and axis

$$r \sin \theta \sim \frac{\sin \theta}{f(\theta)}. \tag{3.18}$$

Minimizing this distance yields an argument minimum of θ_0, where $\cos \theta_0 = d$, i.e. all streamtube throats lie on a cone with opening semi-angle $\theta_0 = \arccos d$, whose axis coincides with the symmetry axis and whose apex is at the origin. This gives a physical interpretation of the number d. The larger the value of d gets, the faster and narrower the fluid stream is. This has a well understood reason: In the next theorem, we will state, that the Landau solutions do not satisfy (2.21) over the origin, but there is a force acting at the origin by which the whole flow is driven. As it turns out, the force and the number d are uniquely related, and a higher number value for d corresponds to a stronger force, entraining the fluid away from the origin. When $|d|$ is near 1 and $\theta \gg \theta_0$, we have that the radial component of u behaves more and more like $u^{(r)} \approx -\frac{2}{r}$. We might consider this the asymptotical inward flow, necessary to compensate the fluid driven away from the origin.

Remark. The assertions of the theorem, in particular the formulae (3.12), and the foregoing remark remain in principle valid, if we do not assume the axis of symmetry to be the x_3-axis, but to be in direction of a nontrivial vector b, called the Landau parameter, and denoting the angle between b and the position vector by θ. Of course, the solution transforms differently from this coordinates to cartesian coordinates then.

Theorem 3.1.2. *For each Landau parameter $b \in \mathbb{R}^3$ there exists a unique weak solution in the sense of distributions on $\mathbb{R}^3$ to the Navier-Stokes system (2.21) with force $b\delta$ which is axisymmetric, but does not* *rotate about $\mathbb{R}\boldsymbol{b}$ and fulfils the homogeneity conditions $u(x) = \frac{1}{|x|}u(\frac{x}{|x|})$ as well as $p(x) = \frac{1}{|x|^2}p(\frac{x}{|x|})$. This solution is a Landau solution, and the magnitude of $\boldsymbol{b}$ and the parameter d are uniquely related. In fact, if β denotes the magnitude of $\boldsymbol{b}$,*

$$\beta(d) = \frac{8\pi}{3d^2(1-d^2)}\left(2d^3 + 6d - 3(1-d^2)\log\frac{1+d}{1-d}\right).$$

If $\boldsymbol{b} = 0$, we understand $(u,p) = (0,0)$ as the solution with $d = 0$.

It should be understood, that a negative sign for the magnitude of the Landau parameter is to be interpreted as a force in the opposite direction of the vector with respect to which θ is defined.

Proof. Due to the properties of u, there exists a continuously differentiable $\tilde{u} : [0,\pi] \to \mathbb{R}^3$, such that $u(x) = r^{-1}\tilde{u}(\theta)$. Then

$$\nabla u^{(r)}(x) = -\frac{1}{r^2}\left(\tilde{u}^{(r)}(\theta)\mathbf{e}_r - \frac{\partial \tilde{u}^{(r)}}{\partial\theta}\mathbf{e}_\theta\right)$$

$$\nabla u^{(\theta)}(x) = -\frac{1}{r^2}\left(\tilde{u}^{(\theta)}(\theta)\mathbf{e}_r - \frac{\partial \tilde{u}^{(\theta)}}{\partial\theta}\mathbf{e}_\theta\right)$$

Since $|\tilde{u}(\theta)|$, $\left|\tilde{u}^{(r)}(\theta)\mathbf{e}_r - \frac{\partial \tilde{u}^{(r)}}{\partial\theta}\mathbf{e}_\theta\right|$ and $\left|\tilde{u}^{(\theta)}(\theta)\mathbf{e}_r - \frac{\partial \tilde{u}^{(\theta)}}{\partial\theta}\mathbf{e}_\theta\right|$ are continuous functions on a compact set, there exist constants C_1, C_2 and C_3, such that for $1 \le p < \frac{3}{2}$

$$\int_{B_\varepsilon(0)} |u(x)|^p\,dV = 2\pi\int_0^\varepsilon\int_0^\pi r^{2-p}|\tilde{u}(\theta)|^p \sin\theta\,d\theta dr \le C_1\int_0^\varepsilon r^{2-p}\,dr < \infty, \tag{3.19}$$

$$\int_{B_\varepsilon(0)} |\nabla u^{(r)}(x)|^p\,dV = 2\pi\int_0^\varepsilon\int_0^\pi r^{2(1-p)}\left|\tilde{u}^{(r)}(\theta)\mathbf{e}_r - \frac{\partial \tilde{u}^{(r)}}{\partial\theta}\mathbf{e}_\theta\right|^p \sin\theta\,d\theta dr \le C_2\int_0^\varepsilon r^{2(1-p)}dr < \infty, \tag{3.20}$$

$$\int_{B_\varepsilon(0)} |\nabla u^{(\theta)}(x)|^p\,dV = 2\pi\int_0^\varepsilon\int_0^\pi r^{2(1-p)}\left|\tilde{u}^{(\theta)}(\theta)\mathbf{e}_r - \frac{\partial \tilde{u}^{(\theta)}}{\partial\theta}\mathbf{e}_\theta\right|^p \sin\theta\,d\theta dr \le C_3\int_0^\varepsilon r^{2(1-p)}dr < \infty. \tag{3.21}$$

Hence, as mentioned but not calculated by M. Cannone and G. Karch [4], one can consider u to be in $W^{1,p}_{loc}(\mathbb{R}^3)$. If $\varphi \in C_0^\infty(\mathbb{R}^3)$ we have thus

$$\langle \operatorname{div} u, \varphi\rangle = -\sum_{i=1}^3 \left\langle u^{(i)}, \frac{\partial\varphi}{\partial x_i}\right\rangle = -\int_{\mathbb{R}^3} u\cdot\nabla\varphi\,dV = \int_{\mathbb{R}^3}\varphi\,\operatorname{div}u\,dV = 0, \tag{3.22}$$

since $\operatorname{div}u \equiv 0$ almost everywhere. That proves $\operatorname{div}u = 0$ in the sense of distributions.

For the statement about the force we first note that u is in $L^2_{loc}(\mathbb{R}^3)$ by (3.19), and thus we can give the formal equation the following distributional sense.

$$\int_{\mathbb{R}^3} (\nabla u - (u\otimes u) - p\mathbb{I})\,\nabla\varphi\,dV = \boldsymbol{b}\varphi(0). \tag{3.23}$$

Without loss of generality, we may assume b to be parallel to the x_3-axis, i.e. $b = \beta e_3$, which simplifies our calculations a lot. Now, if we only integrate over all $|x| \geq \varepsilon$, with $\varepsilon > 0$, we get by integration by parts

$$\int_{\mathbb{R}^3 \setminus B_\varepsilon} (\nabla u - (u \otimes u) - p\mathbb{I}) \, \nabla \varphi \, dV$$

$$= \int_{\mathbb{R}^3 \setminus B_\varepsilon} (-\Delta u + (u \cdot \nabla)u + \nabla p) \, \varphi \, dV \tag{3.24}$$

$$+ \int_{\partial B_\varepsilon} (\nabla u - (u \otimes u) - p\mathbb{I}) \, n(x)\varphi \, dS,$$

where $n(x) = -\frac{x}{\varepsilon}$ is the inward normal to the sphere of radius ε. As seen in theorem (3.1.1), the volume integral on the right-hand side vanishes. For the surface integral we will integrate each component seperately, that is, for $k \in \{1, 2, 3\}$ we consider

$$-\int_{\partial B_\varepsilon} \left((\nabla u_k - u_k u) \cdot \frac{x}{\varepsilon} - p\frac{x_k}{\varepsilon} \right) \varphi \, dS = -\int_{\partial B_1} \varepsilon^2 \left((\nabla u_k - u_k u) \cdot y - p y_k \right) \varphi(\varepsilon y)\frac{1}{\varepsilon^2} \, dS. \tag{3.25}$$

Here, we used the substitution $x = \varepsilon y$ and the fact, that all of the terms $\nabla u_k, u_k u$ and p are (-2)-homogeneous functions. Since $|\varphi| < M$ for some $M \in \mathbb{R}$ and the other terms are in $L^1_{loc}(\mathbb{R}^3)$, we get by the Dominated Convergence Theorem, that

$$\lim_{\varepsilon \to 0+} \int_{\partial B_\varepsilon} \left((\nabla u_k - u_k u) \cdot \frac{x}{\varepsilon} - p\frac{x_k}{\varepsilon} \right) \varphi \, dS = \varphi(0) \int_{\partial B_1} \left((\nabla u_k - u_k u) \cdot x - p x_k \right) dS. \tag{3.26}$$

Now, from (1.1) we derive

$$p x_1 = 2u_1(x), \quad p x_2 = 2u_2(x), \quad p x_3 = 2u_3(x) - \frac{4d}{|x| - d x_3}, \tag{3.27}$$

and furthermore the Euler theorem on homogeneous functions gives us $x \cdot \nabla u_k = -u_k$. So the surface integral on the right-hand side of (3.26) equals for $k = 1, 2$

$$\int_{\partial B_1} u_k + u_k(u \cdot x) + 2u_k \, dS = 0, \tag{3.28}$$

as u_k is an odd function with respect to x_k and $u \cdot x$ is an even function, which means, that the integrand is an odd function and consequently its integral vanishes. So the force is parallel to the axis of rotation. What is left to show is the magnitude of b. First observe, that on ∂B_1

$$u_3(1, \theta) = 2d \frac{1 - 2d \cos \theta + \cos^2 \theta}{(1 - d \cos \theta)^2}, \quad u(x) \cdot x = u^{(r)}(1, \theta) = 2 \left(\frac{1 - d^2}{(1 - d \cos \theta)^2} - 1 \right), \tag{3.29}$$

which can be derived from (3.12) and (3.16) by some elementary calculations. So we compute

$$\int_{\partial B_1} u_3 + u_3(u \cdot x) + 2u_3 - \frac{4d}{1 - d e_3 \cdot x} \, dS$$

$$= -2\pi \int_0^\pi \left(2d \frac{1 - 2d \cos \theta + \cos^2 \theta}{(1 - d \cos \theta)^2} \left(2\frac{1 - d^2}{(1 - d \cos \theta)^2} + 1 \right) \right) \sin \theta \, d\theta - 2\pi \int_{-1}^1 \frac{4d}{1 - d x_3} \, dx_3 \tag{3.30}$$

$$= 4\pi d \int_{-1}^1 \frac{1 - 2d x_3 + x_3^2}{(1 - d x_3)^2} \left(2\frac{1 - d^2}{(1 - d x_3)^2} + 1 \right) dx_3 - 8\pi \log \frac{1 + d}{1 - d}.$$

One checks, that

$$i(x_3) = \frac{5 - 9dx_3 - d^2 + d^3(6x_3 + 6d^2x_3 + 9x_3^3 - d(4 + 9x_3^2 + 3x_3^4))}{3d^3(1 - dx_3)^3} - 2\frac{(d^2 - 1)\log(1 - dx_3)}{d^3} \tag{3.31}$$

is a primitive of the integrand in the last line of (3.30) and that

$$i(1) - i(-1) = 2\frac{2d^3 + 6d - 3(1 - d^2)^2 \log\frac{1+d}{1-d}}{3d^3(1 - d^2)}. \tag{3.32}$$

Combining this result with (3.26), (3.28) and (3.30), we see, that (3.23) holds true.
Since $\beta : (-1, 1) \to \mathbb{R}$ is unbounded, continuously differentiable and its derivative strictly positive, it is bijective. So there exists an inverse function $d : \mathbb{R} \to (-1, 1)$, which proves the claim about the unique relation between β and d. □

Remark. It is not hard to see, that an equivalent formula for β is given by

$$\beta(\theta_0) = \frac{8\pi}{3}\left(8\frac{\cos\theta_0}{\sin^2\theta_0} + \frac{6}{\cos\theta_0} + 3\cos^2\theta_0 \log\left(\frac{1 - \cos\theta_0}{1 + \cos\theta_0}\right)\right). \tag{3.33}$$

In this formula, the first term is the one corresponding to the flux of momentum across a sphere around the origin, that is

$$\left|\int_{\partial B_R}(u \otimes u)\mathbf{n}\,dS\right| = \frac{8\pi}{3}\left(8\frac{\cos\theta_0}{\sin^2\theta_0}\right), \tag{3.34}$$

as can be seen by revising the calculation above. From figure 3.2 it can be seen, that for strong forces this yields the leading term. This term asymptotically equals $\frac{64\pi}{3}\frac{1}{\theta_0^2}$. Now, from this we may give a reasonable estimate for β for a real jet outflowing a small orifice. If we measure the flux of momentum through a closed surface enclosing the origin, we can easily compute θ_0 and therefore β.

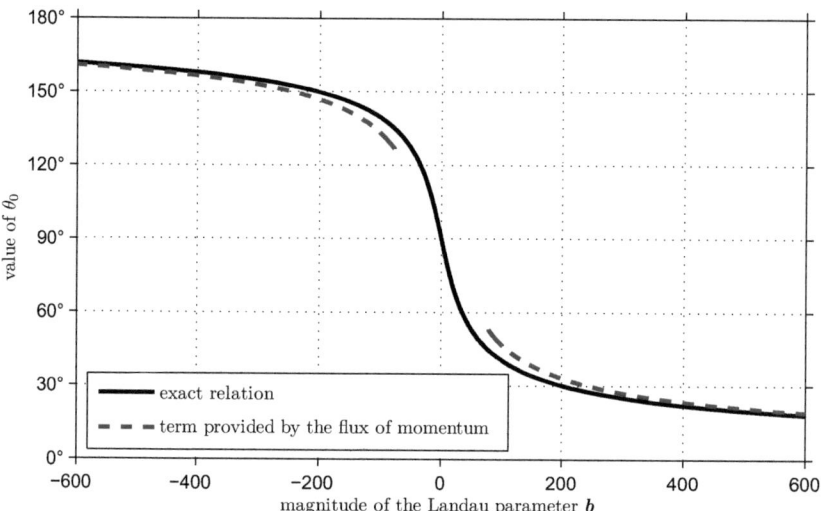

Figure 3.2.: A stronger force produces a lower semi-angle θ_0. The sign of the magnitude indicates the direction of the force with respect to the coordinate system

3.2 Characterisation and approximation of solutions by the Landau solution

In this section, we will give some theorems concerning the analysis of classical Landau solutions, i.e. for Landau solutions with $|d| < 1$. The first theorem is due to V. Šverák [20] and is essential for many applications.

Theorem 3.2.1. *If one drops the symmetry requirement in theorem 3.1.1, it still holds true that every smooth solution to this system, which is (-1)-homogeneous in u and (-2)-homogeneous in p is a Landau solution.*

It is of some interest, that the proof of this theorem uses a quite different approach than the one of theorem 3.1.1. It connects the Landau solutions with the group of conformal, that is angle-preserving, transformations of the unit sphere ∂B_1. We give here a short sketch of the proof.

Sketch of proof. Since the solution is completely determined by its values on the unit sphere, one may decompose u for each point on the sphere into a tangent part v and a normal part $f \mathbf{n}$ and then write down the Navier-Stokes equations as a system of partial differential equations on the sphere in v, f and the pressure p, using the standard conventions for differential operators of Riemannian geometry. Verifying, that $dv = 0$, we see, that there exists a smooth φ on the sphere with $\nabla \varphi = v$. The Navier-Stokes equations now give $-\Delta\mu + \operatorname{div}(\nabla\varphi\mu) = 0$, where $\mu = -\Delta\varphi + 2$. Solving this equation for μ we get $-\Delta\varphi + 2 = 2e^{\varphi}$, which means by [5] (by setting for the there appearing $w = \varphi/2$), that the canonical metric $\bar{g}$ on the sphere and the metric $g := e^{\varphi}\bar{g}$ are isometric. Thus, for a given conformal diffeomorphism h we have

$$\varphi(x) = \log\left|h'(x)\right|^2, \tag{3.35}$$

with $h'(x)$ being the complex derivative of h at x. As shown by [6], all h can be considered to be of the form $h_\lambda = P^{-1} \circ M_\lambda \circ P$ composed with isometries of the sphere, where P is the standard stereographic projection and $M_\lambda(z) = \lambda z$. For φ as in (3.35), composing h with an isometry does either not change φ at all or only changes it by being shifted by this isometry. In any case we can find a coordinate frame such that all solutions of φ take the same form as solutions generated by a h_λ. Now calculating h_λ in the spherical coordinates and setting $\lambda = e^{-\kappa}$, where $\tanh\kappa = d$, we get

$$v = \frac{\partial\varphi}{\partial\theta}e_\theta = \frac{-2d\sin\theta}{1 - d\cos\theta}e_\theta = u^{(\theta)}(1,\theta)e_\theta, \tag{3.36}$$

$$f = -\Delta\varphi = 2\left(\frac{1 - d^2}{(1 - d\cos\theta)^2} - 1\right) = u^{(r)}(1,\theta). \tag{3.37}$$

As $\mathbf{n} = e_r$ on the sphere, the formulae agree for $|x| = 1$ and thus everywhere. □

Given this theorem, it is natural to ask, if one can characterize or at least approximate solutions to the Navier-Stokes equations by Landau solutions under less assumptions. Recent results about this are by A. Korolev and V. Šverák [13], and by H. Miura and T.P. Tsai [16]. For now, let us denote by U^b and P^b the Landau solution with Landau parameter b from theorem 3.1.2.
Korolev and Šverák have been examining solutions to the Navier-Stokes equations near infinity. They proved that assuming some decay condition for the solution, its leading term is given by a Landau solution.

Theorem 3.2.2. *For each $\alpha \in (1, 2)$ there exists $\varepsilon = \varepsilon(\alpha) > 0$ such that the following statement holds true. Let (u,p) be a classical solution of (2.21) on $\mathbb{R}^3\setminus\bar{B}_R$ for some $R > 0$ with zero force satisfying*

$$|u(x)| \le \frac{C_*}{R + |x|} \tag{3.38}$$

with $C_* \leq \varepsilon$. Let $\mathbf{b} = \int_{\partial B_{R_1}} (u \otimes u - T)\mathbf{n}\, dS$ with $R_1 > R$. Then

$$u(x) = U^b(x) + O(|x|^{-\alpha}) \qquad \text{as } |x| \to \infty \tag{3.39}$$

$$p(x) - p_0 = P^b(x) + O(|x|^{-\alpha-1}) \quad \text{as } |x| \to \infty, \tag{3.40}$$

where p_0 is a suitable constant.

A corresponding theorem, which characterizes the behaviour of the solution near the origin, is given for very weak solutions. H. Miura and T.-P. Tsai have shown the following.

Theorem 3.2.3. *For any $q \in (1,3)$, there is a small $C_* > 0$ depending on q such that, if u is a very weak solution of the Navier-Stokes equations with zero force and the property $|u(x)| \leq C_*|x|^{-1}$ in the punctured ball $B_2 \backslash \{0\}$, then there exist a constant C independent of q and u and a scalar function p satisfying $|p(x)| \leq C|x|^{-2}$, unique up to a constant, so that (u,p) satisfies (2.21) with force $\mathbf{b}\delta$ in B_2, where $\mathbf{b} = \int_{\partial B_1} (u \otimes u - T)\mathbf{n}\, dS$. Furthermore*

$$\|u - U^b\|_{W^{1,q}(B_1)} + \sup_{x \in B_1} |x|^{\frac{3}{q}-1} |(u - U^b)(x)| \leq C C_*. \tag{3.41}$$

With this theorem, one can extend the result of Šverák, that is theorem (3.2.1), in the following way. Let (u,p) be a solution of the Navier-Stokes system on $B_2 \backslash \{0\}$ and define for $\lambda > 0$

$$(u_\lambda(x), p_\lambda(x)) = (\lambda u(\lambda x), \lambda^2 p(\lambda x)). \tag{3.42}$$

If there is a $\lambda_1 \in (0,1)$ with $u_{\lambda_1} = u$, then u is called *discretely self-similar*.

Corollary. *If u is discretely self-similar on $B_2 \backslash \{0\}$ and satisfies the assumptions of theorem (3.2.3), then $u \equiv U^b$.*

This is clear in view of the inequality given in the theorem controlling the remainder $u - U^b$ and the fact, that $(u - U^b)(\lambda_1^k x) = \lambda_1^{-k}(u - U^b)(x)$ for $x \in B_1 \backslash B_{\lambda_1}$. Since every (-1)-homogeneous solution is discretely self-similar, this corollary extends Šverák, as Miura and Tsai put it, for small C_*.

As another interesting application, the Landau solution does not only provide the leading terms in the above manner for the standard steady Navier-Stokes system, but also for some fluid flow around a rotating obstacle. Recently, R. Farwig and T. Hishida [8] found the leading term at infinity of a Navier-Stokes flow in the exterior of a rotating obstacle in the following sense.

Theorem 3.2.4. *Let $\omega = a\mathbf{e}_3$ with $a \in \mathbb{R} \backslash \{0\}$ and $\mathbf{b} = \left(\mathbf{e}_3 \cdot \int_{\partial \Omega} T\mathbf{n}\, dS\right)\mathbf{e}_3$, where $\Omega \subset \mathbb{R}^3$, $0 \notin \Omega$, is an exterior domain with smooth boundary. For each $q_0 \in (\frac{3}{2}, 3)$ there exists a constant $\eta = \eta(q_0) > 0$ such that if u is a smooth solution to*

$$\begin{aligned} \operatorname{div} u &= 0, \\ -\Delta u - (\omega \times x) \cdot \nabla u + \omega \times u + \nabla p + u \cdot \nabla u &= 0, \end{aligned} \tag{3.43}$$

on Ω, which satisfies the boundary conditions

$$u = \omega \times x \quad x \in \partial\Omega, \tag{3.44}$$

$$u \to 0 \qquad |x| \to \infty, \tag{3.45}$$

and fulfils

$$\sup_{x \in \Omega} |x||u(x)| + |\mathbf{b}| \leq \eta, \tag{3.46}$$

then for every $q \in (q_0, 3)$ we have

$$u - U^b|_\Omega \in L^q(\Omega), \quad \|u - U^b\|_{L^q(\Omega)} \leq C(|a|^{-\frac{3}{q}+1} + 1) \tag{3.47}$$

with some $C = C(q) > 0$.

This theorem shows, that $u - U^b$ has a better summability, which suggests for q near the value $\frac{3}{2}$ a pointwise decay $1/|x|^2$ at infinity. Noteworth is, that $\boldsymbol{b} = \boldsymbol{b}'$, where one replaces T by $T - u \otimes u$ in the definition of $\boldsymbol{b}'$, since $\boldsymbol{e}_3 \perp \omega \times x$ and $u|_{\partial\Omega} = \omega \times x$. Since the leading term found in theorem (3.2.2) for $a = 0$ is different, the singular behaviour of the difference $u - U^b$ for $a \to 0$ in the rotating case seems natural.

The foregoing statements suggest, that it is quite reasonable to try to classify the classical solutions to the Navier-Stokes system (2.21) on $\mathbb{R}^3 \backslash \{0\}$ with zero force satisfying the decay condition $|u(x)| \leq C_* |x|^{-1}$ for some $C_* > 0$. Šverák conjectured, that all of these solutions are Landau solutions [20]. This question still remains open, yet it has been answered partially by Miura and Tsai [16]. The following theorem tells us, that for small constants $C_* > 0$ this conjecture holds true.

Theorem 3.2.5. *Let $C_* = C_*(q)$ for $q = 2$ from theorem 3.2.3 and let u be a classical solution of the Navier-Stokes system (2.21) on $\mathbb{R}^3 \backslash \{0\}$ with zero force satisfying $|u(x)| \leq C_* |x|^{-1}$. Then u is a Landau solution.*

Proof. Let $\boldsymbol{b}$ be as in theorem 3.2.3 and $w = u - U^b$. If we write $u_\lambda = \lambda u(\lambda x)$ and $w_\lambda = \lambda w(\lambda x)$ for $\lambda > 0$, we get by the scaling-invariance of U^b, that $u_\lambda = U^b + w_\lambda$. Observe, that u_λ still satisfies the same decay condition as u. Thus, by theorem 3.2.3 we get for $q = 2$

$$\left| w_\lambda(x) \right| \leq C C_* |x|^{-\frac{1}{2}}, \quad |x| < 1. \tag{3.48}$$

Note, that the bound does not depend on λ and that λ is positive. We can thus write

$$|w(y)| = \lambda^{-1} |w_\lambda(\lambda^{-1}(y)| \leq C C_* \lambda^{-1} |\lambda^{-1} y|^{-\frac{1}{2}} = C C_* \lambda^{-\frac{1}{2}} |y|^{-\frac{1}{2}}, \quad |y| < \lambda. \tag{3.49}$$

So for all $\varepsilon > 0$ and each $y \in \mathbb{R}^3 \backslash \{0\}$ there is a $\lambda > 0$, such that

$$|w(y)| \leq C C_* \lambda^{-\frac{1}{2}} |y|^{-\frac{1}{2}} < \varepsilon. \tag{3.50}$$

Hence, for the remainder w it holds true that $w = 0$. Therefore $u = U^b$. $\qquad\square$

4 Analysis of the modified Landau solution

Definition 4.0.1. Let $|d| \geq 1$. Then we call a function $(u, p) : \mathbb{R}^3 \backslash \mathscr{C}_d \to \mathbb{R}^3 \times \mathbb{R}$ corresponding to the formulae (1.1) **modified Landau solution**. Here

$$\mathscr{C}_d := \{(x_1, x_2, x_3) \in \mathbb{R}^3 : x_1^2 + x_2^2 - (d^2 - 1)x_3^2 = 0, d\, x_3 \geq 0\} \tag{4.1}$$

is a cone with apex at the origin and opening semi-angle $\arccos \frac{1}{d} =: \theta_c$. A modified Landau solution with $|d| = 1$ is called **critical Landau solution**, for $|d| > 1$ we use the term **supercritical Landau solution**.

From the previous section it is clear, that a **modified Landau solution** still yields a smooth solution to the Navier-Stokes system (2.21) on $\mathbb{R}^3 \backslash \mathscr{C}_d$. This is not hard to see, if one recalls the proof of theorem 3.1.1. There we arrived at an ordinary differential equation of the form $h^2 - 2h_\xi = 0$, which has also the solution

$$h(\xi) = \frac{2d}{1 - d\xi}$$

on $[-1, 1] \backslash \{\frac{1}{d}\}$ for $|d| \geq 1$. But this means exactly, that away from the critical cone the modified Landau solution solves the Navier-Stokes equations. Thus, it might be interesting to investigate, if such modified Landau solutions can be considered to be solutions of the Navier-Stokes system in the whole space $\mathbb{R}^3$. We will start with an examination of the behaviour of critical Landau solutions.

4.1 The critical Landau solution

For $|d| = 1$ the cone shrinks to the **critical half line** $\mathscr{C}_{\pm 1} := \{(x_1, x_2, x_3) \in \mathbb{R}^3 : x_1 = x_2 = 0, \pm x_3 \geq 0\}$. We will here investigate the case $d = 1$. Clearly, for the case $d = -1$ analogous results hold. Then, for u and p we have outside $\mathscr{C}_1$

$$u^{(r)} = -\frac{2}{r}, \quad u^{(\theta)} = -\frac{2}{r}\left(\frac{\sin\theta}{1 - \cos\theta}\right), \quad p = -\frac{4}{r^2}\left(\frac{1}{1 - \cos\theta}\right). \tag{4.2}$$

Note that the radial component is equal to the asymptotical inward flow, which we computed in the classical case. As can be seen from figure 4.1, this is quite reasonable, since for all $0 < \theta \leq \pi$, the fluid is flowing in. Certainly the behaviour of the fluid at the critical half line has to be investigated, as the velocity tends to infinity and all the streamlines end there, suggesting that $\mathscr{C}_1$ acts as a sink. Therefore an investigation of the divergence is of great interest. Computing

$$\int_{B_R} |u| \, dV = \int_0^{2\pi} \int_0^{\pi} \int_0^{R} \left(\frac{2}{r} \sqrt{1 + \frac{\sin^2 \theta}{(1 - \cos \theta)^2}} \right) r^2 \sin \theta \, dr \, d\theta \, d\phi \tag{4.3}$$

$$= 2\pi R^2 \int_0^{\pi} \sqrt{\sin^2 \theta + \frac{\sin^2 \theta}{(1 - \cos \theta)^2} \sin^2 \theta} \, d\theta \tag{4.4}$$

$$= 2\pi R^2 \int_0^{\pi} \sqrt{\sin^2 \theta + (1 + \cos \theta)^2} \, d\theta \tag{4.5}$$

$$= 2\pi R^2 \int_0^{\pi} \sqrt{(2 + 2\cos \theta)} \, d\theta \tag{4.6}$$

$$= 2\pi R^2 \left. \frac{2\sqrt{2} \sin \theta}{\sqrt{1 + \cos \theta}} \right|_0^{\pi} \qquad \text{note, that } \lim_{\theta \to \pi^-} \frac{\sin \theta}{\sqrt{1 + \cos \theta}} = \sqrt{2} \tag{4.7}$$

$$= 8\pi R^2, \tag{4.8}$$

shows that $u \in L^1_{loc}(\mathbb{R}^3)$, because u is continuous away from $\mathscr{C}_1$ and every compact subset of $\mathbb{R}^3$ is an annulus plus or minus compact sets not containing $\mathscr{C}_1$. So one can give a distributional sense to the expression $\operatorname{div} u$ in $\mathscr{D}'(\mathbb{R}^3)$. As we may want to calculate the divergence, let $\varphi \in C_0^\infty(\mathbb{R})$. Then

$$\langle \operatorname{div} u, \varphi \rangle = -\sum_{i=1}^{3} \left\langle u^{(i)}, \frac{\partial \varphi}{\partial x_i} \right\rangle = -\int_{\mathbb{R}^3} u \cdot \nabla \varphi \, dV \tag{4.9}$$

$$= -\int_0^{2\pi} \int_0^{\pi} \int_0^{\infty} (u \cdot \nabla \varphi) \, r^2 \sin \theta \, dr \, d\theta \, d\phi \tag{4.10}$$

$$= -\int_0^{2\pi} \int_0^{\pi} \int_0^{\infty} \left(u^{(r)} \frac{\partial \varphi}{\partial r} + \frac{u^{(\theta)}}{r} \frac{\partial \varphi}{\partial \theta} + \frac{u^{(\phi)}}{r \sin \theta} \frac{\partial \varphi}{\partial \phi} \right) r^2 \sin \theta \, dr \, d\theta \, d\phi \tag{4.11}$$

$$= 2 \int_0^{2\pi} \int_0^{\pi} \int_0^{\infty} \left(r \sin \theta \frac{\partial \varphi}{\partial r} + (1 + \cos \theta) \frac{\partial \varphi}{\partial \theta} \right) dr \, d\theta \, d\phi \tag{4.12}$$

$$= 2 \left(\int_0^{2\pi} \int_0^{\pi} (r \sin \theta) \varphi |_0^\infty \, d\theta \, d\phi - \int_0^{2\pi} \int_0^{\pi} \int_0^{\infty} \varphi \sin \theta \, dr \, d\theta \, d\phi \right. \tag{4.13}$$

$$\left. + \int_0^{2\pi} \int_0^{\infty} (1 + \cos \theta) \varphi |_0^\pi \, dr \, d\phi + \int_0^{2\pi} \int_0^{\pi} \int_0^{\infty} \varphi \sin \theta \, dr \, d\theta \, d\phi \right) \tag{4.14}$$

$$= -4 \int_0^{2\pi} \int_0^{\infty} \varphi(r, 0, \phi) \, dr \, d\phi. \tag{4.15}$$

Integrating over ϕ, which gives us (due to the fact that for $\theta = 0$ the test function φ is constant in ϕ) only an additional factor of 2π, we see that $\operatorname{div} u = -8\pi\delta_{\mathscr{C}_1}$. Not surprisingly, we thus get $\operatorname{supp}\operatorname{div} u = \mathscr{C}_1$. What makes more trouble, is the expression ∇p, since p is obviously no longer in $L^1_{loc}(\mathbb{R}^3)$, which would put us in the position to give a meaning to this expression of the Navier-Stokes equation in the weak sense. The method of principal value integrals, as introduced for example in [12], page 248, is not appropriate for the symmetry of our problem, because it takes the mean value of the function on a set without a small ball centered at one point, where the function is not integrable, and lets the radius of the ball shrink to zero. In our case, where the problem is on the critical half line $\mathscr{C}_1$, we therefore have to use a set with a symmetry appropriate for this situation. One natural set would be a cone with apex at the origin, which we shrink to the critical half line by decreasing the opening semi-angle θ. Another method could be a cylinder with axis of symmetry being $\mathscr{C}_1$. But it can be readily seen, that both of these definitions of a principal value integral would not give us a finite number for p. In fact, the definition with the cone gives us

$$
\begin{aligned}
\text{p.v.} \int_{B_R} p\, dV &= \lim_{\varepsilon \to 0+} \int_0^{2\pi}\int_\varepsilon^\pi\int_0^R -\frac{4}{r^2}\left(\frac{1}{1-\cos\theta}\right) r^2 \sin\theta\, dr d\theta d\phi \\
&= -8\pi R \lim_{\varepsilon \to 0+} \int_\varepsilon^\pi \frac{\sin\theta}{1-\cos\theta}\, d\theta \\
&= -8\pi R \lim_{\varepsilon \to 0+} \left[\log\sin\frac{\theta}{2}\right]_\varepsilon^\pi \\
&= \infty
\end{aligned}
\tag{4.16}
$$

while for the cylinder we have to switch to cylindrical coordinates first, noting that in these coordinates the pressure has the form

$$
p(\rho,\phi,z) = -\frac{4}{\rho^2+z^2}\left(\frac{1}{1-\frac{z}{\sqrt{\rho^2+z^2}}}\right) = -\frac{4}{\rho^2+z^2-z\sqrt{\rho^2+z^2}}.
\tag{4.17}
$$

Then, for the principal value integral over the compact cylinder K of radius R and height H, we have

$$
\begin{aligned}
\text{p.v.} \int_K p\, dV &= \lim_{\varepsilon \to 0+} \int_\varepsilon^R\int_0^H\int_0^{2\pi} -\frac{4}{\rho^2+z^2-z\sqrt{\rho^2+z^2}}\rho\, d\phi dz d\rho \\
&= -8\pi \lim_{\varepsilon \to 0+} \int_\varepsilon^R\int_0^H \frac{\rho}{\rho^2+z^2-z\sqrt{\rho^2+z^2}}\, dz d\rho \\
&= -8\pi \lim_{\varepsilon \to 0+} \int_\varepsilon^R \left(\frac{H+\sqrt{\rho^2+H^2}}{\rho}-1\right) d\rho \\
&= -8\pi \lim_{\varepsilon \to 0+} \left[\sqrt{H^2+\rho^2}-\rho+H\log\left(\frac{H^2r^2}{2H+2\sqrt{H^2+r^2}}\right)\right]_\varepsilon^R \\
&= \infty.
\end{aligned}
\tag{4.18}
$$

Anyway, it is still possible, that there exists a locally integrable function, smooth away from the critical half line having p as a derivative, such that p would act as a derivative of a distribution. For example, consider the function $\tilde{\mathbf{p}} : \mathbb{R}^3 \to \mathbb{R}^3$, defined via

$$\tilde{\mathbf{p}}(r, \theta, \phi) = -\frac{8 \log \sin \frac{\theta}{2}}{r \sin \theta} \mathbf{e}_\theta. \tag{4.19}$$

Then we have $\operatorname{div} \tilde{\mathbf{p}} = p$ and at the same time $\tilde{\mathbf{p}} \in L^1_{loc}(\mathbb{R}^3)$, because

$$\begin{aligned}
\int_{B_R} |\tilde{\mathbf{p}}| \, dV &= \int_0^{2\pi} \int_0^\pi \int_0^R \left| \frac{8 \log \sin \frac{\theta}{2}}{r \sin \theta} \right| r^2 \sin \theta \, dr d\theta d\phi \\
&= 8 \int_0^{2\pi} \int_0^\pi \int_0^R \left| r \log \sin \frac{\theta}{2} \right| dr d\theta d\phi \\
&= 8\pi R^2 \int_0^\pi \left| \log \sin \frac{\theta}{2} \right| d\theta \\
&= 8\pi^2 R^2 \log 2.
\end{aligned} \tag{4.20}$$

Here, the computation of the last integral is non-trivial and can be found in [11]. Now this means, we can give a distributional sense to the expression ∇p, if we write for $\varphi \in C^\infty(\mathbb{R})$

$$\langle \nabla p, \varphi \rangle := \langle \nabla \operatorname{div} \tilde{\mathbf{p}}, \varphi \rangle \tag{4.21}$$

Unfortunately, besides the problem of computing this distribution, not all of our problems are solved. We would like to have $u \in L^2_{loc}(\mathbb{R}^3)$ in order to give a a meaning to $u \cdot \nabla u$. But the computation

$$\begin{aligned}
\int_{B_R} |u|^2 \, dV &= \lim_{\varepsilon \to 0+} 8\pi R \int_\varepsilon^\pi \left(\frac{\sin^2 \theta + (1 + \cos \theta)^2}{\sin \theta} \right) d\theta \\
&= \lim_{\varepsilon \to 0+} 8\pi R \int_\varepsilon^\pi \left(\frac{2 + 2\cos \theta}{\sin \theta} \right) d\theta \\
&= \lim_{\varepsilon \to 0+} 8\pi R \left(4 \log \left(\sin \frac{\theta}{2} \right) \right) \Big|_\varepsilon^\pi \\
&= \infty,
\end{aligned} \tag{4.22}$$

based on (4.8), shows, that this is not the case. Reasoning in the same manner as for the pressure, we see that the symmetry properties of principal value integrals again do not agree with the symmetries in our case.

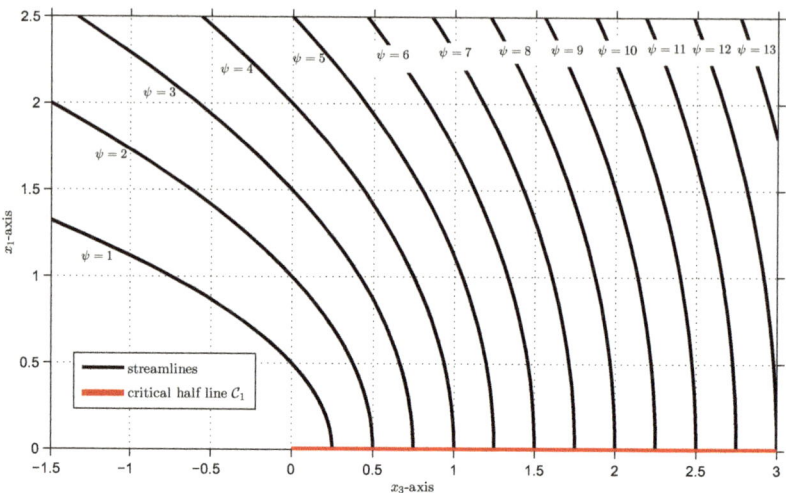

Figure 4.1.: Streamlines for the critical Landau solution. By axisymmetry we see that streamlines corresponding to the same value of the stream function ψ meet on the axis of symmetry.

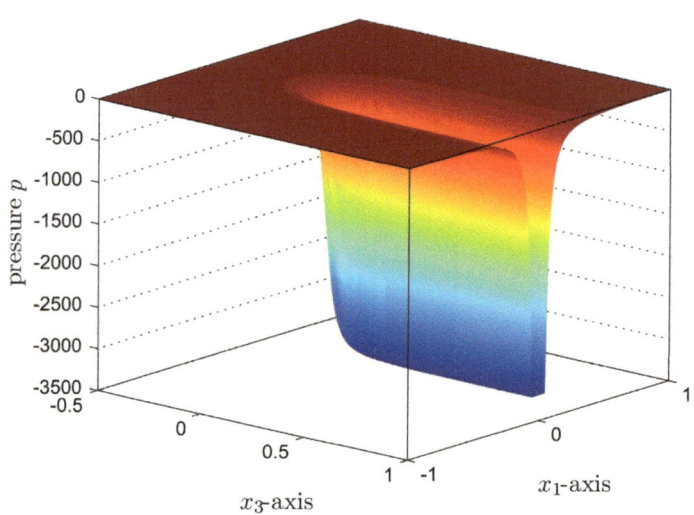

Figure 4.2.: Pressure of the critical Landau solution ($d = 1$). The pressure is not in $L^1_{loc}(\mathbb{R}^3)$, as the decay towards the critical half line is too fast.

4.2 The supercritical Landau solution

In the supercritical case we lose even the property $u \in L^1_{loc}$. For example, for the radial component $u^{(r)}$ we have

$$\frac{1}{2\pi} \int\limits_{B_1(0)} |u^{(r)}| \, dV$$

$$= 2 \left(\lim_{t \to \theta_c-} \int\limits_0^1 \int\limits_0^t \left| \frac{1-d^2}{(1-d\cos\theta)^2} - 1 \right| r\sin\theta \, d\theta dr + \lim_{t \to \theta_c+} \int\limits_0^1 \int\limits_t^\pi \left| \frac{1-d^2}{(1-d\cos\theta)^2} - 1 \right| r\sin\theta \, d\theta dr \right)$$

$$= \left(\lim_{t \to \theta_c-} \int\limits_0^t \left| \left(\frac{1-d^2}{(1-d\cos\theta)^2} - 1 \right) \sin\theta \right| d\theta + \lim_{t \to \theta_c+} \int\limits_t^\pi \left| \left(\frac{1-d^2}{(1-d\cos\theta)^2} - 1 \right) \sin\theta \right| d\theta \right)$$

$$= \left(\lim_{t \to \theta_c-} \left[\left| \frac{1-d^2+(d\cos\theta-1)^2}{d(d\cos\theta-1)} \right| \right]_0^t + \lim_{t \to \theta_c+} \left[\left| \frac{1-d^2+(d\cos\theta-1)^2}{d(d\cos\theta-1)} \right| \right]_t^\pi \right)$$

$$= \infty.$$

Here we used the fact, that the sine is non-negative between 0 and π. In fact, there is no hope to get the integral over $|u|$ convergent, even if one admits principle value integrals. This means there is no way we can give meaning to the formal expression $\operatorname{div} u$ or any other derivative of the solution. We recall the fact, that the stream function $\psi = rf(\theta)$ is constant on a streamline. Since r is non-negative, f cannot change its sign on a stream line. Exactly this happens at the critical angle θ_c. As one can see easily from the definition of f as well as from figure 4.3, each stream line is tending towards the origin for $\theta \to \theta_c$. This certainly suggests that there is no physical interpretation of u as a solution of the Navier-Stokes system, as all stream lines "meet" in one point.

4.3 Summary and outlook

This paper shows that there is no physical interpretation of modified Landau solutions as solutions to the Navier-Stokes system in the whole space, neither in terms of classical, nor weak nor very weak sense. Nevertheless, for the modified Landau solutions we lose uniqueness of the solution on $\mathbb{R}^3 \backslash \mathscr{C}_d$. This is because in the ordinary differential equation (3.10) the constants do not have to be zero, as the solution is not smooth across the critical cone, anyway. So one might want to investigate, if there are other smooth solutions to (2.21), which are axisymmetric about the x_3-axis and (-1)-homogeneous in u as well as (-2)-homogeneous in p. If there is a positive answer to this question, how does it differ from the modified Landau solution and is it reasonably extendable to the whole space?

An examination of the equation (3.10) has been done for example by V. G. Vyskrebtsov [21]. He used a slightly different substitution h defined via

$$f = -(1-\xi)^2 \frac{h_\xi}{h}, \tag{4.23}$$

which leads to the linear equation for h, having the form

$$2\left(1-\xi^2\right)^2 h_{\xi\xi} - \frac{(c_1\xi^2 + c_2\xi + c_3)h}{4} = 0. \tag{4.24}$$

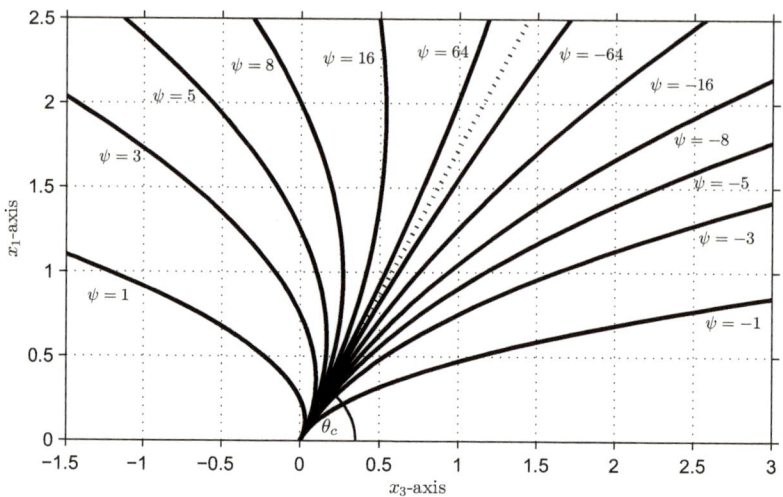

Figure 4.3.: Streamlines of the flow of a supercritical Landau solution for $d = 2$

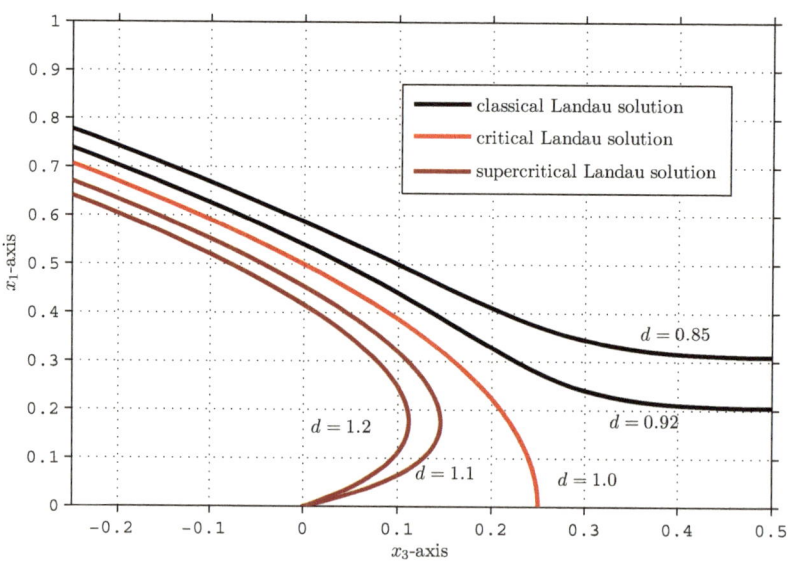

Figure 4.4.: Streamlines of flows for different values near $d = 1$. One can see the change of the behaviour of the flow, when the Landau solution becomes critical or supercritical. Here, $\psi = 1$.

He showed that this equation has an infinite number of linearly independent solutions, when not considering any boundary conditions. The simplest of them is having the form

$$f(\xi) = \gamma(m + \xi), \tag{4.25}$$

where γ and m are arbitrary constants and for the constants of integration we have

$$c_1 = \gamma(\gamma - 2), \quad c_2 = 2m(\gamma^2 - 2\gamma), \quad c_3 = \gamma(\gamma m^2 - 2). \tag{4.26}$$

Note, that the case $\gamma = 2$, $m = 1$ corresponds to the critical Landau solution. A streamline of the flow with constant $\psi = 1$ corresponding to (4.25) is thus defined via the relation

$$r = \frac{1}{f(\theta)} = \frac{1}{\gamma(m + \cos\theta)}. \tag{4.27}$$

This relation reads in terms of cartesian coordinates

$$\sqrt{x_1^2 + x_2^2 + x_3^2} = \frac{1}{\gamma\left(m + \frac{x_3}{\sqrt{x_1^2+x_2^2+x_3^2}}\right)}, \tag{4.28}$$

and thus

$$\sqrt{x_1^2 + x_3^2} + \frac{x_3}{m} = \frac{1}{\gamma m}, \tag{4.29}$$

if one assumes $x_2 = 0$. For $m^2 = 1$, (4.29) describes a parabola of the form

$$x_1^2 = \frac{2}{\gamma m}\left(\frac{\gamma m}{2} - x_3\right). \tag{4.30}$$

For $m^2 \neq 1$, the relation (4.29) can be rewritten as

$$\frac{\left[x_3 + \frac{1}{\gamma(m^2-1)}\right]^2}{m^2/(m^2-1)^2} + \gamma^2(m^2 - 1)x_1^2 = 1. \tag{4.31}$$

If $m^2 > 1$, this equation describes an ellipse, and for $m^2 < 1$, it corresponds to a hyperbola intersecting the axis of symmetry. Because in every case the streamlines intersect this axis, we expect, that the divergence does not vanish there.
So it might be interesting to investigate, for which choices of γ and m these solutions can be reasonably extended to the whole space.

A Appendix

A.1 Spherical Coordinates

We will give here a short overview over the spherical coordinates. Most of the relations and calculations presented here can be found in standard text books, such as Acheson [1], Schade and Neemann [18] or Batchelor [3].

When talking about vector fields such as the velocity $u(x)$, we often find it convenient to glue the coordinate system to the position vector, which we barely notice when working in cartesian coordinates, since the unit vectors e_1, e_2 and e_3 are constant. But we can also represent each vector by its distance from the origin r and two additional angles θ and ϕ, where θ is called the zenith angle and runs from 0 to π and ϕ is the azimuthal angle, running from 0 to 2π. Usually, one chooses θ to be the angle between the positive x_3-axis and the vector, whereas ϕ is the angle between the positive x_1-axis and the orthogonal projection of the vector onto the x_1-x_2-plane. In that case, the spherical coordinates transform into cartesian coordinates as follows

$$
\begin{aligned}
x_1 &= r\sin\theta\cos\phi, \\
x_2 &= r\sin\theta\sin\phi, \\
x_3 &= r\cos\theta.
\end{aligned}
\tag{A.1}
$$

We can now see, that the orthonormal unit vectors of the spherical coordinates are given by

$$
\begin{aligned}
e_r &= \sin\theta\cos\phi\, e_1 + \sin\theta\sin\phi\, e_2 + \cos\theta\, e_3, \\
e_\theta &= \cos\theta\cos\phi\, e_1 + \cos\theta\sin\phi\, e_2 - \sin\theta\, e_3, \\
e_\phi &= -\sin\phi\, e_1 + \cos\phi\, e_2\,.
\end{aligned}
\tag{A.2}
$$

In spherical coordinates, the volume element is given by $dV = r^2\sin\theta\, dr d\theta d\phi$. Similarly, the surface element on a sphere of radius R is $dS = R^2\sin\theta\, d\theta d\phi$. Consequently, if we are integrating a function f, which is axisymmetric about the x_3-axis, we have the formulae (see [12], page A-19)

$$
\int_{B_R} f(x)\, dV = 2\pi \int_0^\pi \int_0^R f(r,\theta) r^2 \sin\theta\, d\theta,
\tag{A.3}
$$

$$
\int_{\partial B_R} f(x)\, dS = 2\pi R^2 \int_0^\pi f(\theta)\sin\theta\, d\theta.
\tag{A.4}
$$

Also, the differential operators transform differently into spherical coordinates. For the gradient, divergence and curl we get

$$
\nabla f = \frac{\partial f}{\partial r} e_r + \frac{1}{r}\frac{\partial f}{\partial\theta} e_\theta + \frac{1}{r\sin\theta}\frac{\partial f}{\partial\phi} e_\phi,
\tag{A.5}
$$

$$
\nabla\cdot a = \frac{1}{r^2}\frac{\partial}{\partial r}\left(r^2 a^{(r)}\right) + \frac{1}{r\sin\theta}\frac{\partial}{\partial\theta}\left(\sin\theta a^{(\theta)}\right) + \frac{1}{r\sin\theta}\frac{\partial a^{(\phi)}}{\partial\phi},
\tag{A.6}
$$

$$
\begin{aligned}
\nabla\times a = {}& \frac{1}{r\sin\theta}\left(\frac{\partial}{\partial\theta}\left(a^{(\phi)}\sin\theta\right) - \frac{\partial a^{(\theta)}}{\partial\phi}\right) e_r \\
&+ \frac{1}{r}\left(\frac{1}{\sin\theta}\frac{\partial a^{(r)}}{\partial\phi} - \frac{\partial}{\partial r}\left(r a^{(\phi)}\right)\right) e_\theta + \frac{1}{r}\left(\frac{\partial}{\partial r}\left(r a^{(\theta)}\right) - \frac{\partial a^{(r)}}{\partial\theta}\right) e_\phi,
\end{aligned}
\tag{A.7}
$$

and for the Laplacian operator we arrive at

$$\Delta f = \frac{1}{r^2}\frac{\partial}{\partial r}\left(r^2\frac{\partial f}{\partial r}\right) + \frac{1}{r^2\sin\theta}\frac{\partial}{\partial\theta}\left(\sin\theta\frac{\partial f}{\partial\theta}\right) + \frac{1}{r^2\sin^2\theta}\frac{\partial^2 f}{\partial\phi^2}. \tag{A.8}$$

With these relations we can formulate the steady Navier-Stokes equation

$$-\Delta u + \nabla p + u\cdot\nabla u = 0 \tag{A.9}$$

in spherical coordinates, where the Laplacian operator appearing in this equation is to be understood in the vectorial sense

$$\Delta u = \nabla(\nabla\cdot u) - \nabla\times(\nabla\times u). \tag{A.10}$$

In view of the main part of this paper, we here assume $u^{(\phi)} = 0$ and $\frac{\partial}{\partial\phi} = 0$, which results in only two equations for $u^{(r)}$ and $u^{(\theta)}$

$$u^{(r)}\frac{\partial u^{(r)}}{\partial r} + \frac{u^{(\theta)}}{r}\frac{\partial u^{(r)}}{\partial\theta} + \frac{\left(u^{(\theta)}\right)^2}{r} = -\frac{\partial p}{\partial r} + \frac{1}{r^2}\frac{\partial}{\partial r}\left(r^2\frac{\partial u^{(r)}}{\partial r}\right)$$
$$+ \frac{1}{r^2\sin\theta}\frac{\partial}{\partial\theta}\left(\sin\theta\frac{\partial u^{(r)}}{\partial\theta}\right) - \frac{2u^{(r)}}{r^2} - \frac{2}{r^2}\frac{\partial u^{(\theta)}}{\partial\theta} - \frac{2u^{(\theta)}\cot\theta}{r^2} \tag{A.11}$$

$$u^{(r)}\frac{\partial u^{(\theta)}}{\partial r} + \frac{u^{(\theta)}}{r}\frac{\partial u^{(\theta)}}{\partial\theta} + \frac{u^{(r)}u^{(v)}}{r} = -\frac{1}{r}\frac{\partial p}{\partial\theta} + \frac{1}{r^2}\frac{\partial}{\partial r}\left(r^2\frac{\partial u^{(\theta)}}{\partial r}\right)$$
$$+ \frac{1}{r^2\sin\theta}\frac{\partial}{\partial\theta}\left(\sin\theta\frac{\partial u^{(\theta)}}{\partial\theta}\right) + \frac{2}{r^2}\frac{\partial u^{(r)}}{\partial\theta} - \frac{u^{(\theta)}}{r^2\sin^2\theta} \tag{A.12}$$

A.2 Euler Theorem on Homogeneous Functions

Another well-known theorem is the Euler Theorem on Homogeneous Functions, which can be found for example in [2]. A function $f : V \to W$ between to two $\mathbb{R}$-vector spaces is said to be α-homogeneous, if for all $\lambda > 0$ and all $x \in V$ we have $f(\lambda x) = \lambda^\alpha f(x)$.

Euler Theorem on Homogeneous Functions. *Let $f : \mathbb{R}^n\backslash\{0\} \to \mathbb{R}$ be continuously differentiable. Then f is α-homogeneous if and only if for all $x \in \mathbb{R}^n\backslash\{0\}$ it holds true, that $x\cdot\nabla f(x) = \alpha f(x)$.*

Proof. This can be seen by looking at

$$x\cdot\nabla f(x) = \frac{d}{d\lambda}\left(f(\lambda x)\right)\Big|_{\lambda=1} = \frac{d}{d\lambda}\left(\lambda^\alpha f(x)\right)\Big|_{\lambda=1} = \alpha f(x). \tag{A.13}$$

The converse holds in the exact manner by integrating. $\qquad\square$

Corollary. *If f is continuously differentiable and α-homogeneous, then all of its partial derivatives are $(\alpha - 1)$-homogeneous.*

Proof. One simply checks, that

$$\lambda\frac{\partial}{\partial x_i}f(x) = \frac{\partial}{\partial x_i}f(\lambda x) = \lambda^\alpha\frac{\partial}{\partial x_i}f(x). \tag{A.14}$$

$\qquad\square$

A.3 The Stereographic Projection

In the proof of Šverák's theorem 3.2.1 we make use of the stereographic projection, which we will introduce in this section. The definition here is based on [12], page A-22, yet slightly adjusted to the proof of the theorem. Let $S \subset \mathbb{R}^3$ be the two dimensional unit sphere and define the stereographic projection $P : S \backslash \{e_3\} \to \mathbb{C}$ by

$$P(x_1, x_2, x_3) = \frac{x_1}{1 - x_3} + i \frac{x_2}{1 - x_3}. \tag{A.15}$$

Then P is bijective and for the inverse map we have (where we write $z = a + ib$)

$$P^{-1}(z) = \left(\frac{2a}{1 + |z|^2}, \frac{2b}{1 + |z|^2}, \frac{|z|^2 - 1}{1 + |z|^2} \right). \tag{A.16}$$

If we switch to the spherical coordinates on the sphere (θ, ϕ), the formulae transform to

$$P(\theta, \phi) = \cot \frac{\theta}{2} \cdot e^{i\phi}, \tag{A.17}$$

$$P^{-1}(z) = \left(2 \arctan \frac{1}{|z|}, \varphi \right), \tag{A.18}$$

where $z = |z| e^{i\varphi}$ and $P^{-1}(0) = (\pi, 0)$. The stereographic projection is conformal, i.e. it preserves angles of intersecting curves.

Then if we want to calculate the map $h_\lambda = P^{-1} \circ M_\lambda \circ P$, where $M_\lambda : \mathbb{C} \to \mathbb{C}$ is given by $M_\lambda(z) = \lambda z$, we get

$$h_\lambda(\theta, \phi) = \left(2 \arctan \frac{\tan \frac{\theta}{2}}{|\lambda|}, \phi \right) \tag{A.19}$$

A.4 Distributions

In this paper, we often use the notion of distributions. Distributions are elements of the dual space $\mathscr{D}'(\Omega)$ of the space of test functions $\mathscr{D}(\Omega) = C_0^\infty(\Omega)$. That is, a distribution $f : \mathscr{D}(\Omega) \to \mathbb{R}$ takes a test function φ to the number $\langle f, \varphi \rangle$, where the operation $\langle f, \cdot \rangle$ is linear and continuous in the following sense: If a sequence of test functions φ_n is converging to φ in the space of test functions (that is, there exists a compact subset of Ω containing the support of every φ_n and φ, and φ_n and derivatives of φ_n of arbitrary order converge uniformly to those of φ), then $\langle f, \varphi_n \rangle$ is converging to $\langle f, \varphi \rangle$ in $\mathbb{R}$.

One of the most famous examples of a distribution is the **Dirac distribution** $\delta : \mathscr{D}(\mathbb{R}) \to \mathbb{R}$, which is defined via

$$\langle \delta, \varphi \rangle = \varphi(0). \tag{A.20}$$

It is easy to check, that δ is indeed in the dual space of $\mathscr{D}(\mathbb{R})$. Furthermore, for $f \in L^1_{loc}(\Omega)$ we can regard f to be a distribution by setting

$$\langle f, \varphi \rangle = \int_\Omega f \varphi \, dV. \tag{A.21}$$

Note, that this mapping is well-defined: In fact we are only integrating over the compact set $\operatorname{supp}\varphi$, and so we see that

$$\langle f, \varphi \rangle = \int_\Omega f\varphi\, dV \tag{A.22}$$

$$= \int_{\operatorname{supp}\varphi} f\varphi\, dV \tag{A.23}$$

$$\leq \|\varphi\|_\infty \int_{\operatorname{supp}\varphi} f\, dV \tag{A.24}$$

$$\leq C \|\varphi\|_\infty, \tag{A.25}$$

where for given f the constant C does only depend on $\operatorname{supp}\varphi$. By the Lebesgue dominated convergence theorem we see, that $\langle f, \varphi_n \rangle$ really converges to $\langle f, \varphi \rangle$. A distribution, which has a representation in terms of a integral operator as in (A.21) is called regular. Obviously, the Dirac distribution is not regular. As $\mathscr{D}(G)$ has a natural embedding into $\mathscr{D}(\Omega)$ for any open G contained in Ω, a distribution on $\mathscr{D}(\Omega)$ defines a distribution on $\mathscr{D}(G)$ by restriction. We say, that a distribution vanishes on an open set G, if for all $\varphi \in \mathscr{D}(G)$ it holds that $\langle f, \varphi \rangle = 0$. Observe, that a distribution vanishes on G, if it vanishes on a neighborhood of every point in G. That makes sure, that if a distribution vanishes on each member of a family of sets, then it also vanishes on the union. Thus, for a distribution f on $\mathscr{D}(\Omega)$, the set

$$N_f = \bigcup \{ G \subset \Omega : G \text{ is open and } f \text{ vanishes on } G \} \tag{A.26}$$

is the maximal open set, on which f vanishes. The complement of this set $\mathbb{R}^n \backslash N_f$ is called the support of f. Observe, that for a regular distribution, the classical definition of the support of a function and the support of a distribution coincide.

Let us finally introduce the concept of derivation in the space of distributions. For $f \in \mathscr{D}'(\Omega)$, we define the derivative of f with respect to x_j as

$$\left\langle \frac{\partial f}{\partial x_j}, \varphi \right\rangle := -\left\langle f, \frac{\partial \varphi}{\partial x_j} \right\rangle. \tag{A.27}$$

The derivative of f with respect to x_j is thus a distribution itself. If we generalize this definition to arbitrary derivatives by setting

$$\langle D^\alpha f, \varphi \rangle := (-1)^{|\alpha|} \langle f, D^\alpha \varphi \rangle \tag{A.28}$$

we can say, that every distribution is infinitely often differentiable.

Bibliography

[1] D. J. Acheson. *Elementary fluid dynamics*. Oxford Applied Mathematics and Computing Science Series. The Clarendon Press Oxford University Press, New York, 1990.

[2] H. Amann and J. Escher. *Analysis II*. Grundstudium Mathematik. [Basic Study of Mathematics]. Birkhäuser Verlag, Basel, 1999.

[3] G. K. Batchelor. *An Introduction to Fluid Dynamics*. Cambridge University Press, 1967.

[4] M. Cannone and G. Karch. Smooth or singular solutions to the Navier-Stokes system? *J. Differential Equations*, 197:247–274, 2004.

[5] S.-Y. A. Chang and P. C. Yang. The inequality of Moser and Trudinger and applications to conformal geometry. *Communications on Pure and Applied Mathematics*, Vol. LVI:1135–1150, 2003.

[6] B. A. Dubrovin, A. T. Fomenko, and S. P. Novikov. *Modern geometry—methods and applications. Part I*, volume 93 of *Graduate Texts in Mathematics*. Springer-Verlag, New York, second edition, 1992.

[7] R. Farwig and T. Hishida. Stationary Navier-Stokes Flow Around a Rotating Obstacle. *Funkcialaj Ekvacioj*, 50(3):371–403, 2007.

[8] R. Farwig and T. Hishida. Asymptotic Profile of Steady Stokes Flow around a Rotating Obstacle. *FB Mathematik, Technische Universität Darmstadt*, preprint no. 2578 (2009). manuskripta mathematica (to appear 2010).

[9] M. Feistauer. *Mathematical methods in fluid dynamics*, volume 67 of *Pitman Monographs and Surveys in Pure and Applied Mathematics*. Longman Scientific & Technical, Harlow, 1993.

[10] J. Felcman. Matematické modelování ve fyzice 1. Lecture notes, *Matematicko-fyzikální fakulta, Univerzita Karlova v Praze*, 2009.

[11] I.S. Gradshteyn and I.M. Ryzhik. *Table of Integrals, Series and Products*. Academic Press, Orlando, seventh edition, 2007.

[12] L. Grafakos. *Classical and Modern Fourier Analysis*. Pearson Education, Inc., New Jersey, 2004.

[13] A. Korolev and V. Šverák. On the large-distance asymptotics of steady state solutions of the Navier-Stokes equations in 3D exterior domains. *arXiv:math/07110560*, 2007.

[14] J. Kurzweil. *Obyčejné diferenciální rovnice*. Státní Nakladatelství Technické Literatury (SNTL), Prague, 1978.

[15] L. D. Landau. A new exact solution of Navier-Stokes equations. *Doklady Acad. Sci. URSS*, 43:286–288, 1944.

[16] Hideyuki Miura and Tai-Peng Tsai. Point singularities of 3D stationary Navier-Stokes flow. *arXiv:0810.2004v2*, 2009.

[17] M. Renardy and R. C. Rogers. *An introduction to partial differential equations*. Springer-Verlag, New York, second edition, 2004.

[18] H. Schade and K. Neemann. *Tensoranalysis*. de Gruyter Lehrbuch, Berlin, third revised edition, 2009.

[19] H. B. Squire. The round laminar jet. *The Quarterly Journal of Mechanics and Applied Mathematics*, 4:321–329, 1951.

[20] V. Šverák. On Landau's solutions of the Navier-Stokes equations. *arXiv:math/0604550*, 2006.

[21] V. G. Vyskrebtsov. New exact solutions of the Navier-Stokes equations for axisymmetric automodel flows of fluids. *Journal of Mathematical Sciences*, 104:1456–1463, 2001.